Planet Earth Is Conscious-And Life Exists in Amazing Places

This book explores the concepts of our planet Earth being a conscious entity, and all of the incredible places life has been found, around, on, and inside our planet.

I believe that planet Earth (also known as Gaia) is a conscious being in its own right. Many of the ancients also believed the same thing.

Even metaphysical philosophies like Theosophy talk about the Gaia consciousness and that there are even higher level beings like the Solar Logos.

There is an amazing amount of life on Earth—much of which has only been discovered in recent decades.

For example: did you know that life penetrates the Earth's crust down to at least seven miles beneath the Ocean's seabed, and possibly much deeper.

That life exists in Lake Vostok beneath the Antarctic Ice which hasn't been in contact with the surface for many thousands of years?

There has also been new bacteria found on the International Space Station and in meteorites which have hit the Earth.

Life exists on our planet in many unusual and amazing places and you will be amazed at its diversity.

Planet Earth Is Conscious-And Life Exists in Amazing Places

Planet Earth Is Conscious-And Life Exists in Amazing Places

Copyright Page

This book is copyrighted for 2022

Title Planet Earth Is Conscious

Subtitle: And Life Exists in Amazing Places

By Martin K. Ettington

All Rights Reserved USA 2022

ISBN: 9798431428623

Printed in the United States of America

Planet Earth Is Conscious-And Life Exists in Amazing Places

Planet Earth Is Conscious-And Life Exists in Amazing Places

Other books by Martin K. Ettington

The God Like Powers Series:

Spiritual and Metaphysics Books:
Prophecy: A History and How to Guide
God Like Powers and Abilities
Enlightenment for Newbies
Removing Illusions to Find True Happiness
Using the Scientific Method to Study the Paranormal
A Compendium of Metaphysics and How to Guides (Six books together in one volume)
Love from the Heart
The Enlightenment Experience
Learn Your Soul's Purpose
Pursuing Enlightenment
A Modern Man's Search for Truth
Use Intuition and Prophecy to Improve Your Life
The Handbook of Spiritual and Energy Healing
Pure Spirituality and God
Memories Before Birth and Reincarnation
Druid History, Mysticism, Rituals, Magic, and Prophecy

Longevity & Immortality:
Physical Immortality: A History and How to Guide
The Commentaries of Living Immortals
Records of Extremely Long Lived Persons
Enlightenment and Immortality
Longevity Improvements from Science
The 10 Principles of Personal Longevity
Telomeres & Longevity
The Diets and Lifestyles of the World's Oldest Peoples
The Longevity Six Books Bundle
Long Lived Plants and Animals

Science Fiction:
Out of This Universe
The Immortals of the Interstellar Colony
The Mystic Soldier
The Immortality Sci Fi Bundle
Visiting Many Universes

The History of Science Fiction and Fantasy

Human Invisibility
Invulnerability and Shielding

Teleportation
Psychokinesis
Our Energy Body, Auras, and Thoughtforms
The God Like Powers Series—Volume 1 Compilation

The Yoga Discovery Series:
Yoga-An Ancient Art Form
Hatha Yoga-Helping you Live Better
Raja Yoga-Through the Ages
The Yoga Discovery Package

Business & Coaching Books:
Creating, Paublishing, & Marketing Practitioner Ebooks
Building a Successful Longevity Coaching Business
Why Become a Coach?
The Professional Coaching Success Trilogy
2020-Make Money Writing and Selling Books
The 2020 Handbook of High Paying Work Without a College Degree
The important of Creativity and How to Improve Yours
Quantum Mechanics, Technology, Consciousness, and the Multiverse

Self-Improvement
Stress Relief and Methods to do so
The Importance of Creativity and How to Improve Yours
Building Self-Confidence
See the World Clearly
A Trilogy of Self Help Books

Science, Technology, and Misc.
Future Predictions By and Engineer & Seer
The Unusual Science & Technology Bundle
The Real Atlantis-In the Eye of the Sahara
Removing Limits On Our Consciousness-And Thinking Outside the Box
Universal Holistic Philosophy
Ball Lightning

Planet Earth Is Conscious-And Life Exists in Amazing Places

Stranger Than Science Stories and Facts

Survival
Survival of Humanity Throughout the Ages
33 Incredible True Survival Stories
The Importance of Fire in History and Mythology
How to Survive Anything: From the Wilderness to Man Made Disasters
Building and Stocking a Nuclear Shelter for less than $10,000
The Human Survival Five Books Bundle

Legendary Beings
Are Cryptozoological Animals Real or Imaginary?
Fire in History and Mythology
All About Dragons
Sea Serpents and Ocean Monsters
The Legendary Animals Five Books Bundle
The Mythical People of Ireland
Bigfoot Mysteries and Some Answers
About the Little People: Fairies, Elves, Dwarfs and Leprechauns

Ancient History
The Real Atlantis-In the Eye of the Sahara
Ancient & Prehistoric Civilizations
Ancient & Prehistoric Civilizations-Book Two
The History of Antediluvian Giants
The Antediluvian History of Earth
Ancient Underground Cities and Tunnels
Strange Objects Which Should Not Exist
More Out of Place Artifacts
Strange and Ancient Places in the USA
A Theory of Ancient Prehistory And Giant Aliens
The Destruction of Civilization About 10,500 B.C.
A Timeline of Intelligent Life on Earth
A 300 Million Year Old Civilization Existed on Earth

Aliens and Space
Aliens and Secret Technology
Aliens Are Already Among Us
Designing and Building Space Colonies
Humanity and the Universe
All About Moon Bases
All About Mars Journeys and Settlement
The Space and Aliens Six Books Bundle
A Theory of Ancient Prehistory and Giant Aliens
The Space Colonies and Space Structures Coloring Book
All About Asteroids
Spaceships, Past, Present, and Future
Astronauts, Cosmonauts, and Other Important Space Flyers
All About Mars Journeys and Settlement
Mining the Asteroid Belt

Time Travel and Dimensions
Real Time Travel Stories From a Psychic Engineer
The Real Nature of Time: An Analysis of Physics, Prophecy, and Time Travel Experiences
Stories of Parallel Dimensions
We Live in a Malleable Reality-and We Can Change It
The Time, Dimensions, and Quantum Mechanical Bundle
Alternate Dimensions & the Otherworld

Political and Social

The Empire of the United States: Forged By God's Spirit Through Man

Planet Earth Is Conscious-And Life Exists in Amazing Places

<u>The Longevity Training Series</u>

(A transcription of the online Multimedia Longevity Coaching Training Program)

The Personal Longevity Training Series-Book1-Long Lived Persons
The Personal Longevity Training Series-Book2-Your Soul's Purpose
The Personal Longevity Training Series-Book3-Enable Your Life Urge
The Personal Longevity Training Series-Book4-Your Spiritual Connection
The Personal Longevity Training Series-Book5-Having Love in Your Heart
The Personal Longevity Training Series-Book6-Energy Body Health
The Personal Longevity Training Series-Book7-The Science of Longevity
The Personal Longevity Training Series-Book8-Physical Body Health
The Personal Longevity Training Series-Book9-Avoiding Accidents
The Personal Longevity Training Series-Book10-Implementing These Principles

The Personal Longevity Training Series-Books One Thru Ten

These books are all available in digital and printed formats from my website and on Amazon, Barnes & Noble, Apple ITunes, and many other sites

My Books Website is: <u>http://mkettingtonbooks.com</u>

Planet Earth Is Conscious-And Life Exists in Amazing Places

Signup for our Mailing List to get the following:

1) A discount coupon for 25% discount on all books on our site
2) Occasional Notices of new books available
3) Occasional Email on other offerings of ours (Monthly)

If you have any questions about this book or other subjects please contact the Author at:

mke@mkettingtonbooks.com

Planet Earth Is Conscious-And Life Exists in Amazing Places

Table of Contents

1.0 Introduction 1
2.0 The Gaia Hypothesis 3
3.0 Gaia Consciousness..5
4.0 Life Deep in the Earth 9
5.0 Life in the Ocean...17
   5.1 Volcanic Vents 17
   5.2 Weird Life in the Oceans....................................27
6.0 Life Deep in the Jungles 37
7.0 Life Beneath and in the Antarctica Oceans 67
   7.1 Life under the Ice 67
   7.2 Lake Vostok under Miles of Ice 71
8.0 Life in the Atmosphere..75
9.0 Life in Isolation...79
   9.1 Tepui-Plateaus in South America............................79
   9.2 Life in Deep Caves 85
10.0 Life in Space 89
   10.1 Bacteria on the ISS 89
   10.2 Life in Meteorites 97
11.0 Summary 101
12.0 Bibliography..103

Planet Earth Is Conscious-And Life Exists in Amazing Places

Planet Earth Is Conscious-And Life Exists in Amazing Places

1.0 Introduction

This book explores the concepts of our planet Earth being a conscious entity, and all of the incredible places life has been found, around, on, and inside our planet.

I believe that planet Earth (also known as Gaia) is a conscious being in its own right. Many of the ancients also believed the same thing.

Even metaphysical philosophies like Theosophy talk about the Gaia consciousness and that there are even higher level beings like the Solar Logos.

There is an amazing amount of life on Earth—much of which has only been discovered in recent decades.

For example: did you know that life penetrates the Earth's crust down to at least seven miles beneath the Ocean's seabed, and possibly much deeper.

That life exists in Lake Vostok beneath the Antarctic Ice which hasn't been in contact with the surface for many thousands of years?

There has also been new bacteria found on the International Space Station and in meteorites which have hit the Earth.

Life exists on our planet in many unusual and amazing places and you will be amazed at its diversity.

Planet Earth Is Conscious-And Life Exists in Amazing Places

2.0 The Gaia Hypothesis

The study of planetary habitability is partly based upon extrapolation from knowledge of the Earth's conditions, as the Earth is the only planet currently known to harbour life. The Gaia hypothesis, also known as Gaia theory or Gaia principle, proposes that all organisms and their inorganic surroundings on Earth are closely integrated to form a Single and self-regulating complex system, maintaining the conditions for life on the planet.

The scientific investigation of the Gaia hypothesis focuses on observing how the biosphere and the evolution of life forms contribute to the stability of global temperature, ocean salinity, oxygen in the atmosphere and other factors of habitability in a preferred homeostasis.

The Gaia hypothesis was formulated by the chemist James Lovelock and co-developed by the microbiologist Lynn Margulis in the 1970s. Initially received with hostility by the scientific community, it is now studied in the disciplines of geophysiology and Earth system science, and some of its principles have been adopted in fields like biogeochemistry and systems ecology. This ecological hypothesis has also inspired analogies and various interpretations in social sciences, politics, and religion under a vague philosophy and movement.

Overview

The Gaia theory posits that the Earth is a self-regulating complex system involving the biosphere, the atmosphere, the hydrospheres and the pedosphere, tightly coupled as an evolving system. The theory sustains that this system as a whole, called Gaia, seeks a physical and chemical environment optimal for contemporary life.

Planet Earth Is Conscious-And Life Exists in Amazing Places

Gaia evolves through a cybernetic feedback system operated unconsciously by the biota, leading to broad stabilization of the conditions of habitability in a full homeostasis. Many processes in the Earth's surface essential for the conditions of life depend on the interaction of living forms, especially microorganisms, with inorganic elements.

These processes establish a global control system that regulates Earth's surface temperature, atmosphere composition and ocean salinity, powered by the global thermodynamic desequilibrium state of the Earth system.

The existence of a planetary homeostasis influenced by living forms had been observed previously in the field of biogeochemistry, and it is being investigated also in other fields like Earth system science. The originality of the Gaia theory relies on the assessment that such homeostatic balance is actively pursued with the goal of keeping the optimal conditions for life, even when terrestrial or external events menace them.

3.0 Gaia Consciousness

In the 1970s, the chemist James Lovelock proposed the Gaia Hypothesis, named after the Greek Earth Goddess, Gaia. The Gaia Hypothesis proposed that life on Earth is a self-regulating system involving the biosphere, the atmosphere, the hydrosphere, and the pedosphere (skin of soil and living organisms), all of which are intimately integrated as an evolving complex system. Wider research proved the original hypothesis wrong, in the sense that it is not organic life alone but the whole Earth system that is self-regulating.

However, the hypothesis has been modified and elaborated enough and there have been enough predictions made and confirmed that the Gaia Hypothesis has become the Gaia Theory, which now holds that the Earth system as a whole seeks a physical and chemical environment optimal for contemporary life.

Acceptance of the Gaia Theory has become so widespread that, in 2001, a thousand scientists at the European Geophysical Union meeting signed the Declaration of Amsterdam, starting with the statement *"The Earth System behaves as a single, self-regulating system with physical, chemical, biological, and human components.*

Originally, many Earth scientists strongly criticized the Gaia Hypothesis, suggesting, among other things, that it involved a teleological explanation (rather than the type of mechanico/deterministic explanation favored by traditional science).

Lovelock responded that "Nowhere in our writings do we express the idea that planetary self-regulation is purposeful, or involves foresight or planning by the biota."

The Earth Is Alive

However, maybe Lovelock hasn't gone far enough. It seems possible to simplify and strengthen the Gaia Theory if we take the strong form of the Anthropic Principle seriously, which asserts that the universe must be compatible with conscious life, which means that it is most likely that the universe is a living place that contains a spectrum of consciousness.

From this perspective, we can reformulate the Gaia Theory as "The Earth is alive and behaves in a purposive fashion with a type of consciousness in order to pursue an environment optimal for life."

The proposed consciousness of the Earth would clearly be located at a place on the consciousness spectrum that is different from ordinary human consciousness or human consciousness as augmented by the technologies of traditional science. But there is significant evidence that people who train their consciousness, e.g. Yogis and Taoists in the Eastern metaphysical tradition, can reach and participate in wider portions of the consciousness spectrum, and, in some cases, touch the unique form of consciousness that is manifested by the Earth.

However, the Eastern maps that have emerged from this participation have tended to lack the useful corrective of systematic external observation (the central purview of Western science) and they, by and large, have not provided us with a future oriented philosophy powerful enough to orient and guide us through the great transformation that is taking place right now.

However, if it is true that it is possible to connect with the consciousness of the Earth, then we may be able to find ways to partner with the Earth in accomplishing the One

Planet Earth Is Conscious-And Life Exists in Amazing Places

Purpose of Planetary Philosophy, i.e. the evolution of a higher order living system within which the human species takes its place furthering and being furthered by the life of the Earth. The question then becomes, how do we do this?

Planet Earth Is Conscious-And Life Exists in Amazing Places

4.0 Life Deep in the Earth

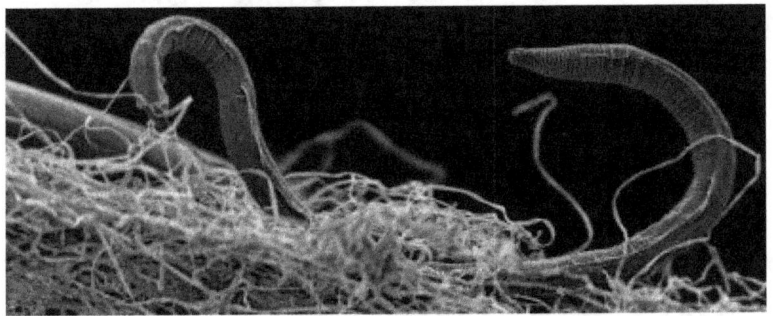

Scientists estimate that subterranean organisms constitute a massive amount of carbon, 245 to 385 times greater than that contained in all humans.

Organisms of Earth's deep underground constitute between 15 billion and 23 billion tons of carbon and occupy an estimated volume almost twice that of the oceans combined, scientists from the Deep Carbon Observatory reported yesterday (December 10) in advance of the annual meeting of the American Geophysical Union in Washington, DC.

The scientific team, which includes hundreds of researchers from all over the world, drilled boreholes kilometers below the continents and seafloor to sample microbes. The information collected by the scientists has allowed them to build models of the deep ecosystem and make the estimates of the deep life biomass.

The researchers found a stunning array of life, mostly microbial, and estimate that approximately 70 percent of the total number of Earth's bacteria and archaea organisms live in this realm. These microbes live at extremes of pressure, temperature, and nutrient and energy availability.

"Exploring the deep subsurface is akin to exploring the Amazon rainforest. There is life everywhere, and everywhere there's an awe-inspiring abundance of unexpected and unusual organisms," says Mitch Sogin, a scientist at the Marine Biological Laboratory and co-chair of the Deep Carbon Observatory's Deep Life team, in a statement.

Many questions remain as to how life spreads under the surface, which energy sources are the most important to sustain these organisms, and whether this is where life began on planet Earth.

The findings also suggest that extraterrestrial life may be similarly hidden underground.

"I think it's probably reasonable to assume that the subsurface of other planets and their moons are habitable, especially since we've seen here on Earth that organisms can function far away from sunlight using the energy provided directly from the rocks deep underground," Rick Colwell, a member of the Deep Carbon Observatory team from Oregon State University, tells *BBC News*.

Another Article about Deep Bioms:

Something odd is stirring in the depths of Canada's Kidd Mine. The zinc and copper mine, 350 miles northwest of Toronto, is the deepest spot ever explored on land and the reservoir of the oldest known water. And yet 7,900 feet below the surface, in perpetual darkness and in waters that have remained undisturbed for up to two billion years, the mine is teeming with life.

Many scientists had doubted that anything could live under such extreme conditions. But in July, a team led by University of Toronto geologist Barbara Sherwood Lollar reported that the mine's dark, deep water harbors a population of remarkable microbes.

The single-celled organisms don't need oxygen because they breathe sulfur compounds. Nor do they need sunlight. Instead, they live off chemicals in the surrounding rock — in particular, the glittery mineral pyrite, commonly known as fool's gold.

"It's a fascinating system where the organisms are literally eating fool's gold to survive," Sherwood Lollar said. "What we are finding is so exciting — like 'being a kid again' level exciting."

Sherwood Lollar is excited not only because of how peculiar the mine's rock-eating life seems, but also because of the growing realization that strange forms of life might not be so peculiar after all. Scientists are starting to find similar microbes in other deep spots, including boreholes, volcanic vents on the bottom of the ocean and buried sediments far beneath the seafloor.

"The deep microbial realm reveals a biosphere that's more extensive, resilient, varied and strange than we had realized," said Robert Hazen, a mineralogist at the Carnegie Institution's Geophysical Laboratory in Washington, and co-founder of Deep Carbon Observatory, a global project to study the deep biosphere.

Cut off from light, air, and any connection to the surface, this shadowy realm seems more like an alien world than part of Earth. Hazen said exploring it could help us understand how life might have begun on other planets as well as on our own. We might even find alien-like creatures living undetected right beneath our feet.

Lots of life at the bottom

Sherwood Lollar's work builds on a 2018 report by Deep Carbon Observatory scientists who tried to map the total extent of Earth's deep biosphere comprehensively for the first time.

In the eye-opening report, a team led by Cara Magnabosco, a geobiologist at the Swiss technical university ETH Zurich, estimated that some 5×10^{29} cells live in the deep Earth: that's five-hundred-thousand-trillion-trillion cells. Collectively, they weigh 300 times as much as all living people combined. The team describes this hidden ecosystem as an "underground Galapagos."

An exterior shot of Kidd Mine.

The denizens of the deep are an exotic bunch even beyond their appetite for solid rock. One species, the microbe *Geogemma barossii,* can live at temperatures of 250 degrees Fahrenheit — well above the boiling point of water and close to the theoretical limit at which vital organic molecules start to disintegrate.

Separate studies of material drilled near the Mariana Trench in the Pacific Ocean hint that some organisms could be living six miles below the seafloor, limited only by the heat at such tremendous depths. Laboratory experiments show that some microbes can tolerate pressures 20,000 times higher than the air pressure at sea level, meaning that there are almost certainly more extreme ecosystems out there than the one in the Kidd Mine.

"We're finding that we don't really understand the limits to life," Sherwood Lollar said.

The pace of existence in the deep also seems radically different from that on the surface. In ancient environments like the trapped waters at the bottom of the Kidd Mine, food and energy are scarce. To compensate, cellular metabolism slows almost to a standstill.

"Many of the microbes may survive for thousands of years or more without dividing, just replacing their broken parts," said Karen Lloyd, a University of Tennessee microbiologist who studies life at the bottom of the ocean.

There are so many deep microbes that, despite a seemingly lazy existence, they collectively exert a huge impact on their habitats. For instance, a community of cells on the ocean floor consume methane gas that bubbles up from ancient sediment. "Deep subsurface microbes eat massive amounts of methane that would otherwise be released," Lloyd said, helping curb atmospheric levels of a potent greenhouse gas.

Back to beginnings

One of the big questions facing Sherwood Lollar is how the deep-life community at the Kidd Mine is related to those found in other mines or stretched out beneath the oceans. "The number of systems we've looked at so far really is limited," she said, "but they probably had a single origin at some point in life's 4-billion-year history."

If so, there should still be clues about when and how life first colonized the deep.

Fossils show that surface life has changed enormously over billions of years, but slow-motion deep life may retain much of its primitive characteristics. That's especially true at the Kidd Mine, which is in one of the oldest, most stable portions of Earth's crust. (The rock in and around the mine have lain undisturbed for 2.7 billion years, and have been cool enough to support life for at least 2 billion years.)

Planet Earth Is Conscious-And Life Exists in Amazing Places

Cara Magnabosco and colleagues collect ancient water samples 4,300 feet deep within the Beatrix Gold Mine, South Africa to investigate the diversity and abundance of deep microbes.

Sherwood Lollar wants to sequence the genes of the Kidd Creek microbes and then do a 23andMe-style analysis to unravel their kinship to other residents of the deep Earth:

Are they all still close relatives, or have they diversified and adapted significantly to their local environments? It's a delicate project, but she hopes to have results within a year or two.

Such studies could offer hints about where life first arose on Earth. Charles Darwin imagined the beginning might have occurred in a warm little pond, but "there's absolutely no reason why it could not have been a warm little rock fracture," Sherwood Lollar said. In many ways, she noted, sulfur-breathing microbes living beneath thick, protective layers of rock would have been well suited to the brutal conditions on our planet when it was young.

Another, even wilder possibility is that life originated more than once, with other forms still surviving somewhere on Earth. "We've literally only scratched the surface of the deep biosphere," Hazen said. "Might there be entire domains that are not dependent on the DNA, RNA and protein basis of life as we know it?" Perhaps we just haven't found them yet.

Paul Davies, a physicist at Arizona State University, has long advocated systematic searches for such "shadow life." The recent forays into the deep biosphere show how it might be done. Since known organisms cannot survive above 250 degrees Fahrenheit, Davies suggests going to extreme environments (around undersea volcanic vents, for instance) and checking for anything that appears alive at temperatures around 300 to 400 degrees Fahrenheit. "That would stand out as a candidate for shadow life," he said.

Ever cautious, Sherwood Lollar points out that she hasn't found any evidence of shadow life at the Kidd Mine. But she heartily agrees that scientists need to keep a wide-open mind about what could be lurking within the deep world: "We see only what we look for. If we don't look for something, we miss it."

5.0 Life in the Ocean

Over 71 percent of the Earth is covered by Oceans and we know very little about what goes on down there. It has only been within recent decades that momentous discoveries were made of life outside volcanic vents in the deep ocean, and many new species of animals are still being discovered every day.

5.1 Volcanic Vents

It was February 1977, and Robert Ballard, a marine geologist at Woods Hole Oceanographic Institution (WHOI), sat aboard the research vessel Knorr 400 miles off the South American coast, staring at photos before him.

"I think there's shimmering water right over here to the left, coming out right off the top."

The photos had been taken by cameras towed 8,000 feet (2,500 meters) below the surface on a platform called

Planet Earth Is Conscious-And Life Exists in Amazing Places

ANGUS. They unveiled a discovery that would turn our understanding of life on Earth on its head: Warm water was drifting out of the seafloor along the Galápagos Rift.

Ballard, along with a team of thirty marine geologists, geochemists, and geophysicists, had found the world's first known active hydrothermal vent. There were no biologists aboard—because no one had expected the second shocking discovery that came soon after: Life was thriving in the abyss. Foot-long clams and human-sized tube worms with tulip-looking heads made the already extraterrestrial landscape look, well, alien.

Hydrothermal vents form in volcanic areas where subseafloor chambers of rising magma create undersea mountain ranges known as mid-ocean ridges. Cold seawater seeps into cracks in the seafloor and can be heated up to a raging 750° F (400° C) by interacting with magma-heated subsurface rocks. The heat stimulates chemical reactions that pull in minerals and chemicals from the rocks, before the fluids percolate back up through vent openings as a chemical-laden soup.

It turns out that nutrients and chemicals belching out of the vents were fueling a rich and productive ecosystem. Communities of microbes fed off chemicals in the vent fluids. The microbes were hosted symbiotically by the strange creatures of the deep, which provided shelter in exchange for food. No plants, no sunlight. Just microbes converting carbon dioxide in the ocean into organic compounds—for themselves and for their hosts.

The long-held notion that life at the bottom of the ocean couldn't exist without food that rained down from the sunlit surface was tossed out the window. Along with photosynthesis, there was chemosynthesis supporting an entirely new kind of ecosystem in the abyss.

"Everyone sat around speechless," said Ballard. "It was like processing a nonlinear equation. It was pretty amazing to find these creatures."

Two years after the first vent was found off Galápagos, scientists exploring another mid-ocean ridge a few hundred miles north found never-before-seen geysers of hot, dark, mineral-rich fluid erupting from tall, chimney like structures jutting up from the seafloor. The fluids trailed away in underwater plumes like smoke from smokestacks. They called these new types of vents black smokers.

Since then, hundreds of vents have been discovered across the global ocean, from Antarctica to the Arctic, along with an estimated eight hundred vent animal species and countless microbial species. The rate of discovery shows no signs of leveling off.

In August, 2017, WHOI scientist Stefan Sievert organized the Elizabeth W. and Henry A. Morss Colloquium "Life Without Sunlight at Deep-Sea Hot Springs" to celebrate the 40th anniversary of the discovery of these unique ecosystems and to inform the public about the implications of chemosynthesis for life on Earth and possibly other planetary bodies. It coincided with the 6th International Symposium on Chemosynthetic-Based Ecosystems (CBE6) at which scientists from around the world convened to discuss the current state of hydrothermal research and where things are headed as our understanding of life without sunlight evolves.

Old vents, new places

As scientists began finding additional vent sites in the decades following the initial discovery, most were in volcanically active areas along mid-ocean ridges similar to the Galápagos Rift. In the early 2000s, however, things began to change. Scientists discovered a vent system known as the Lost City Hydrothermal Field near the ridge axis in the Atlantic Ocean. Sitting on an ancient slab of seafloor crust, the field contained vent structures as tall as the Leaning Tower of Pisa and, unlike previously discovered vents, this system was hosted not in crustal rocks, but in peridotite, the rock type that makes up most of Earth's upper mantle.

"Folks began finding hydrothermal systems that were not hosted in basalt—the main rock we find in the oceanic crust," said Frieder Klein, a geochemist at WHOI. "This had important implications for our understanding of these systems and their associated ecosystems, because it makes a difference whether seawater is reacting with oceanic crust, or with Earth's mantle." It changes the chemistry of fluids emanating from the vents.

As scientists expanded their explorations in other types of geological settings—at the margins of continents and arc island volcanoes, for example, or at subductions zones, where one plate dives beneath another—they found a diversity of vents and other kinds of seafloor fluid flow that can support chemosynthetic life, said WHOI scientist Chris German.

Exploration enablers

Finding vent systems in diverse oceanic environments takes curiosity, determination, and, well, guts. It also takes some pretty robust technology. Metal-crushing pressure, scorching-hot seawater, and rugged, dark landscapes are just some of the extreme conditions that make vent research tough on scientists and the tools they bring down there.

Fortunately, the challenges of extreme deep-sea exploration have led to tight collaboration between marine scientists and engineers and the emergence of a variety of enabling technologies driving these new discoveries.

Towed platforms such as ANGUS and human-occupied submersibles such as Alvin were followed by tethered remotely operated vehicles such as Jason. Then came deep-diving autonomous underwater vehicles, or AUVs. These pilotless vehicles swim at depths of 6,000 meters, or nearly 4 miles, performing a number of key functions, including high-resolution seafloor mapping, collecting seawater data, and imaging.

"Before the arrival of these autonomous vehicles, we could only suspend in situ sensors from deep-tow cables, which made operations very unwieldy," said German. "We had to lower gear to the seafloor on thick cables, which then could only be towed slowly through the oceans at 1 to 1.5 knots and in straight lines to avoid entanglement. With AUVs, we can attach the same sensors and make tight turns and systematically map things out in 3-D grid patterns." That gives scientists the ability to visualize entire vent fields over a 5-mile range.

To work well in the deep sea, an AUV needs the ability to hover, stop, and reverse in unknown terrain, while mapping out various shapes and hazards as it approaches a vent site. And, if it gets stuck, it needs the smarts to bail itself out.

"In 2005, we had ABE, our Autonomous Benthic Explorer, parked nose-up against a black-smoker chimney at three thousand meters in the remote South Atlantic," said German. "After ten minutes, ABE reversed back the way it had come to get dislodged, stepped across ten meters to the left, and got back on with the program."

German also credits developments in sensor technology as a breakthrough area for vent research. In particular, in situ sensors that can find hydrothermal plumes have been key for investigating new sites since the 1980s, when oceanographers began using optical clarity sensors to look for murky, mineral-laden plumes spewing from black-smoker vents. More recently, scientists have also advanced technology for chemical sensors that can detect chemical signals in hydrothermal plumes.

"These sensors offer a great way to prospect for submarine fluids that only recently entered the ocean and haven't yet had a chance to fully react with seawater," said German "and that tells us when we are getting close to a source. The main challenge for the future will be to provide enough power for the sensors, so they can provide reliable and stable data sets for long periods in the deep ocean, as we develop next-generation robots that can explore for days and weeks on end, over hundreds of kilometers."

When it comes to scientists and technology, of course, there's always a wish list. German has his sights set on a sensor that can simultaneously measure a variety of key "tracer" elements in a hydrothermal plume in real time. This way, he and other scientists could know what types of seafloor fluids to expect at a site before a submersible heads down to the seafloor to investigate in detail.

Teeming with life

From the first jaw-dropping glimpses in the late 1970s, we've known that vents are teeming with curious life: jumbo clams, deep-sea tubeworms, Yeti crabs, and shrimp with primitive "eyes" that detect black body radiation emitted from hot objects (such as vents). The shrimp, as well as other vent animals, live in a complex symbiosis with bacteria. Despite the absence of sunlight, all of the essential ingredients are there: heat from the Earth, mineral-rich vent fluids, and a vast universe of microbes that use chemicals produced by these volcanic systems—such as hydrogen sulfide, hydrogen, and even natural gas—as energy sources.

But at what rates are these microorganisms using the chemicals? How much carbon do they produce and how fast do they grow? And what role do they play in supporting the deep sea and beyond? These, according to WHOI biologist Stefan Sievert, are fundamental questions that have been on scientists' minds since vents were first discovered and investigated by WHOI scientists such as Holger Jannasch, Carl Wirsen, and Craig Taylor, among others.

"There's always been a great need to place hydrothermal systems in more of a quantifiable biogeochemical context for the rest of the ocean," he said. "We've known that the mixing of hydrothermal fluid with seawater is driving chemosynthesis, and that there appears to be high activity at vent sites and biological processes happening faster than in the rest of the deep ocean. But we haven't been able to really quantify how fast the microbes are oxidizing chemicals and growing and how much biomass—particularly carbon, the building block of life—they are producing."

Sievert points out that most of our current understanding of vent ecosystem productivity is based on theoretical estimates and lab experiments—not on direct observations at the actual vent sites. This means the pressure, temperatures, and chemical concentrations microbes are exposed to in the lab may not correspond well with what they experience a few thousand meters down.

To bridge this knowledge gap, Jesse McNichol, a former graduate student in Sievert's lab, and other scientists at WHOI and elsewhere have performed incubation experiments on cruises in which they've collected deep-sea vent microbes in water samplers that maintain the seafloor pressure of the vent sites. The scientists then analyze the samples to measure the rates and activities of the microbes.

The research is making headway, thanks to powerful new analytical tools that can, for example, match up a microbe's identity with its biochemical activity down to the level of a single cell. A recently developed instrument known as Vent-SID, for Vent Submersible Incubation Device, enables scientists to incubate microbes and measure their growth rates even right at the seafloor.

"What we've found is that these microbes are really active and quite fast growing," Sievert said. "In fact, based on our shipboard experiments, some can double their numbers within a few hours. That's at least as fast as many of the microbes we find in the surface ocean."

Sievert adds that these techniques are helping scientists to shed more light on interactions among organisms in vent food webs. They are also helping them assess the role of deep-sea vents in cycling chemicals such as carbon, nitrogen, and sulfur between rocks, the ocean, and living things. How much carbon, for example, is recycled within the food web at deep-sea vents versus exported to the surrounding deep ocean?

"Most people are never going to go down to a vent site in a submersible or physically experience one of these ecosystems," she said, "but they can value the existence of it the same way they value the existence of other species on Earth and the stewardship and conservation of those species."

Planet Earth Is Conscious-And Life Exists in Amazing Places

5.2 Weird Life in the Oceans

There are many strange lifeforms in the Oceans. Here are a few of them:

1. Clown FrogFish

This strange creature from waters around Indonesia is bright yellow with red markings. But some frogfish, who like to keep a low profile, can change colors to blend in with their surroundings.

Clown Frogfish sometimes change their colors to blend into their environment.

2. Sea Pen

A sea pen is actually a colony of polyps with tentacles that form branches and look like a quill pen. Sea pens, found in the shallow waters of New Zealand, root to the sea floor with an anchoring bulb to which the colony can retreat if it is threatened.

3. Blob Sculpin

These bulbous creatures are hard to spot because they are reclusive by nature. The two-foot-long fish nest on sea floors up to a mile deep off California's coast, and the males guard the eggs until they hatch.

Blob Sculpin

4. Red-Lipped Batfish

Red lips attract prey and fins help these fish "walk" on the seafloor near the Galapagos Islands. They also have a retractable snout that they can use to get prey close to their mouths.

Red Lip Batfish

5. Giant Spider Crab

Native to Japan, giant spider crabs can grow up to 12 feet long. They forage on the ocean floor up to 1,000 feet deep, which is nearly 2 miles.

Japanese spider crab under water in an aquarium in Japan

6. Giant Tube Worms

These amazing creatures can survive with zero sunlight, freezing temperatures, and the crushing pressure of the deep undersea environment—and some are even adapted to live at the edge of hydrothermal vents in the East Pacific, out of which spurts superheated water infused with toxic chemicals.

7. Vampire Squid

These animals have the largest eyes of any animal proportional to their size. They live at 10,000 feet under the surface in the mesopelagic and bathypelagic regions of the world's temperate and tropical world oceans and can survive in the low-oxygen environment of those layers.

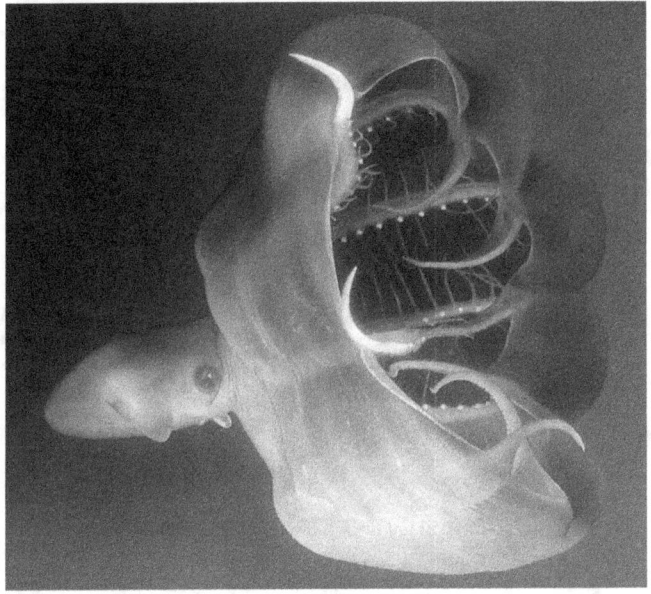

Vampire Squid

8. Leafy Seadragon

The fins of this creature look like leaves and help it move through the water as well as camouflage it to look like drifting seaweed. These animals are found on the southwestern coast of Australia.

Is it a plant or a leafy seadragon?

9. Kiwa, God of Shellfish, Crab

This crab was discovered more recently south of Easter Island on a hydrothermal vent 5,000 feet under the surface. It has fur covering its long claws that contains bacteria to detoxify its food, and it is likely blind.

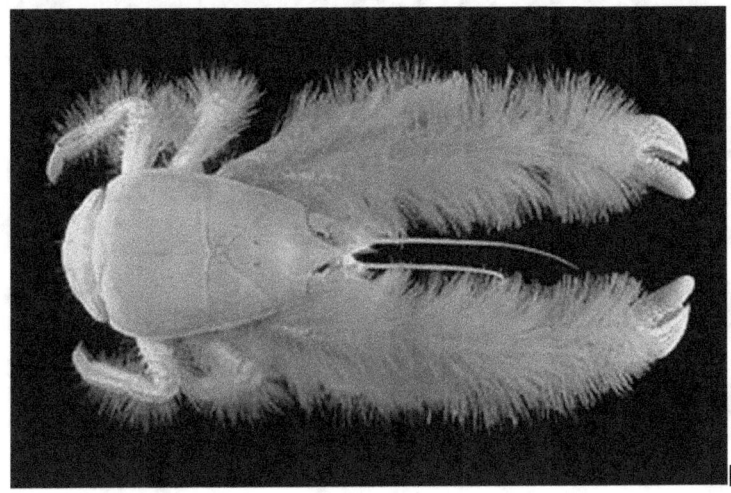

Kiwa

10. Metapseudes

Lots of these odd arthropod creatures were found in coral rubble in Western Australia, but marine biologists aren't sure yet what it does, only that it looks strange.

Metapseudes

6.0 Life Deep in the Jungles

New life is still being discovered deep inside the jungles of the world. Scientists estimate that there are **at least 30,000** as yet undiscovered plants, most of which are rainforest species. The tropics are the earth's richest natural reserves.

1. Amazonian tree with human-sized leaves finally gets ID'd as new species

Coccoloba gigantifolia leaves can reach 2.5 meters (8 feet) in length.

This is a story of incredible patience. Botanists first encountered individuals of this tree in 1982 while surveying the Madeira River Basin in the Brazilian Amazon. They knew it was a species of *Coccoloba*, a genus of flowering plants that grows in the tropical forests of the Americas, but they couldn't pinpoint the species. The individual trees they came across weren't bearing any flowers or fruits then, parts that are essential to describing a plant species, and the trees' leaves were too large to dehydrate, press and carry back with them.

While the plant, and its massive leaves, became locally famous, it was only in 2005, that the researchers finally collected some seeds and dying flowers from a tree. These materials weren't good enough to describe the plant species, but the researchers sowed the seeds at the campus of the National Institute of Amazonian Research (INPA), grew the seedlings, and waited. Thirteen years later, in 2018, one of the planted trees flourished and fruited, finally giving the researchers the botanical material they needed to describe the new species. The new species, named *C. gigantifolia* in reference to the plant's giant leaves, grows to about 15 meters (49 feet) in height and has leaves that can reach 2.5 meters (8 feet) in length, likely the largest known leaf among dicotyledonous plants — a large group of flowering plants that include sunflowers, hibiscus, tomatoes and roses.

2. This grouper species collected from an Australian fish market almost became someone's dinner

Queensland Museum ichthyologist Jeff Johnson with a specimen of Epinephelus fuscomarginatus.

It's not every day that you find a previously undescribed species in a fish market, but that's exactly what Jeff Johnson, an ichthyologist with Australia's Queensland Museum, did. He had first heard of a mystery grouper 15 years back, and since then, received occasional photographs of the fish, which he thought was a potential new-to-science species. In 2017, when a fisherman sent Johnson a picture of the grouper yet again, Johnson tracked down the market where the fisherman had sold the fish, and bought all five individuals of the fish he found there. Then, together with his colleagues, Johnson analyzed the fish's DNA, and compared it to those of related species in the museum. Finally, in a new paper published this year, the researchers confirmed that the grouper is indeed new to science, and they named it *Epinephelus fuscomarginatus*.

3. Meet Mini mum, Mini ature, Mini scule: Tiny new frogs from Madagascar

An adult male Mini mum, one of the world's smallest frogs, rests on a fingernail with room to spare.

This year, herpetologists introduced us to three previously undescribed species of extremely small frogs from Madagascar, aptly named *Mini mum, Mini ature, and Mini scule.* All of them, just a few millimeters long, belong to *Mini*, a genus that is also entirely new to science. The new frog species are known from just a handful of locations, and may already be threatened with extinction. Researchers have recorded *Mini mum* only in Manombo Special Reserve in southeast Madagascar, for example, while *Mini scule* is known only from the fragmented forests of Sainte Luce Reserve. The areas in which the frogs occur are also likely small, threatened and declining.

Planet Earth Is Conscious-And Life Exists in Amazing Places

4. Newly described Chinese giant salamander may be world's largest amphibian

A. sligoi or South China giant salamander painting.

For a long time, the Chinese giant salamander, which reaches lengths of more than 5 feet (1.6 meters) and is the world's largest known living species of amphibian, was considered to be a single species, *Andrias davidianus*. In the past, some researchers did suspect that the salamander was probably multiple species, but a new study published this year backed the suspicion with evidence. Researchers analyzed samples of the salamander from a series of historical museum specimens to see what local wild populations of the amphibian may have been like before humans started farming the animals and moving them around extensively, and found that the salamander is not just one, but three distinct species. These include *A. davidianus*, *A. sligoi*, and a third species that hadn't been named at the time the study was published. Of the three recognized species, the South China giant salamander (*A. sligoi*) is most likely the largest, reaching 2 meters (6 feet) in length, the researchers say.

5. It took 25 years to describe Indonesia's newest tarsier

Niemitz's tarsier from the Togean Islands of Indonesia.

Scientists Alexandra Nietsch and Carsten Niemitz first spotted this tarsier on the Togean Islands off Sulawesi, Indonesia, in 1993. Locals have known of the primate by the names *bunsing*, *tangkasi* and *podi*. But it took researchers more than 25 years of detailed study, including the tarsier's vocalization and DNA, to finally describe the small primate as a species that's new to science in a paper published in 2019. They named the species Niemitz's tarsier (*Tarsius niemitzi*) in honor of the scientist who was one of the first to bring it to the attention of the scientific world. The description of Niemitz's tarsier has increased the number of known tarsier species in Sulawesi and surrounding islands to 12, but the authors say the islands could be home to at least 16.

6. Some of these nine newly described Fijian bees are restricted to a single mountaintop

H omalictus terminalis is found only within 95 meters of Mount Batilamu's peak.

This year, researchers described nine new species of bees from the island country of Fiji in the southern Pacific Ocean. These colorful bees in shades of black, golden-green, and metallic, with hints of purple iridescence, are part of the genus *Homalictus* Cockerell, a group that's not been taxonomically reviewed in Fiji for 40 years. Many of these bees either have very restricted distributions or are known only from single mountaintops, according to the researchers, and could soon become extinct due to changes in climate and other environmental risks. One new-to-science species, *Homalictus terminalis*, for example, has only been found on Mount Batilamu, where it seems to be restricted to the top 95 meters (312 feet) of the mountain peak. Another newly described species, *H. ostridorsum*, has only been recorded on Mount Tomanivi, while *H. taveuni* is named after the island of Taveuni, the only place it is known from.

7. Newly described marmoset species live in part of Amazon forest area that's fast disappearing

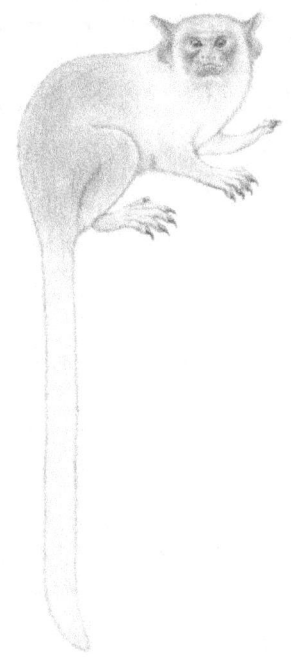

A sketch of Mico munduruku.

When researchers surveying the stretch of Amazon forest lying between the Tapajós and Jamanxim rivers in the Brazilian state of Pará, chanced upon a group of three marmosets with white tails, they suspected that it was a potential new-to-science species. White tails are very rare among primates that live in South America; only one other primate species have it. The researchers were right. After studying the monkeys in both the forest and the laboratory, they confirmed that the marmoset, with its distinct white tail, white forearms with a beige-yellowish spot on the elbow, and white feet and hands, was a new species. The marmoset has been named *Mico munduruku* after the Munduruku, an indigenous group of people who live in the Tapajós–Jamanxim interfluve. It's not all good news, though. The Amazon forest that's home to *M. munduruku* is being rapidly cut for agricultural expansion, logging, hydroelectric power plants, and gold mining.

8. This new-to-science monkey lives in an 'island' amid deforestation in Brazil

Plecturocebus parecis (left) and the closely related Plecturocebus cinerascens (right).

This year, scientists announced a second, new-to-science species of monkey, also found in the Amazon rainforest. The grey monkeys, named *Plecturocebus parecis* after the Parecis plateau in Rondônia in Brazil where they are found, were first seen by scientists in 1914. Locally known as the "otôhô," researchers subsequently saw the monkeys once again in 2011 and confirmed that it was sufficiently distinct from the closely related ashy black titi to be classified as a separate species. The titi monkey's habitat lies within in the "Arc of Deforestation," an area of high deforestation where vast swathes of forest have been cleared for cattle ranching and mechanized soy farms. But so far, the monkeys seem to have escaped some of the damage because the steep slopes of the plateau they occupy offer them protection by making the habitat hard to access and unappealing for large-scale deforestation.

Planet Earth Is Conscious-And Life Exists in Amazing Places

9. New species of orange-red praying mantis mimics a wasp

Vespamantoida wherleyi.

Praying mantises tend to resemble leaves or tree trunks and come in shades of green and brown. But in 2013, researchers spotted a bright orange-red mantis with a black abdomen in a research station on the banks of the Amazon River in northern Peru. The praying mantis was not only mimicking a wasp's bright colors, but also a wasp's movements. Such conspicuous mimicry of wasps is rare among mantises, making this finding exciting, the researchers wrote in a paper published this year. The new-to-science species was named *Vespamantoida wherleyi*, the genus name *Vespamantoida* meaning wasp-mantis.

Planet Earth Is Conscious-And Life Exists in Amazing Places

10. New species of giant flying squirrel brings hope to one of the worlds 'most wanted.'

The Mount Gaoligong flying squirrel, or Biswamoyopterus gaoligongensis, was recently discovered last year in Yunnan, China, by Quan Li of the Kunming Institute of Zoology and his team.

Giant flying squirrels belonging to the group *Biswamoyopterus* are incredibly rare. The first species described in the genus, the Namdapha flying squirrel (*Biswamoyopterus biswasi*) from India, has been seen by researchers only once in 1981. Its Laotian relative, the Laotian giant flying squirrel (Biswamoyopterus laoensis), was first spotted by researchers in bush meat markets of Lao PDR (Laos) in 2012. Both species are known from a single specimen each. This year, scientists in China introduced us to a third species of the genus, Mount Gaoligong flying squirrel (*B. gaoligongensis*), which they first spotted in the collection of the Chinese Academy of Sciences. Thankfully, the team could subsequently observe the animals in the field as well and collect another specimen. Compared to the "lost" Namdapha and Laotian giant flying squirrels, researchers say that Mount Gaoligong flying squirrel's conservation status looks "slightly optimistic."

11. Newly described pocket shark likely glows in the dark

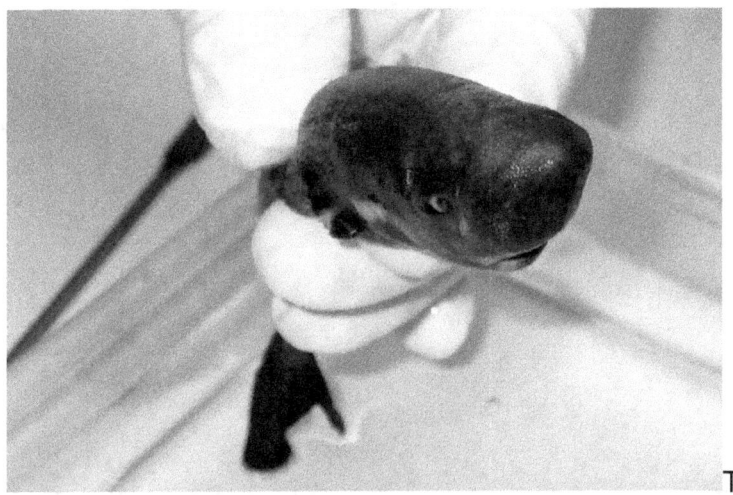

The only known specimen of the American pocket shark was discovered in the Gulf of Mexico in 2010.

The new-to-science American pocket shark (*Mollisquama mississippiensis*) is the world's second pocket shark species to be described. Researchers were surveying the eastern Gulf of Mexico to study what sperm whales eat when they collected a large sample of animals from the ocean's depths. Among the collection was a small shark that the team hadn't seen before. As it turned out, the animal was a previously undescribed species of pocket shark (the pocket shark gets its names not for its small size but because of small pocket-like openings or glands found behind each of its pectoral fins). This year, the researchers introduced the American pocket shark to us in a new paper, noting that the species has numerous light-producing organs or photophores covering much of the body, which possibly helps the shark glow in the dark depths of the deep sea.

Planet Earth Is Conscious-And Life Exists in Amazing Places

12. Newly described tree species from custard apple family is likely endangered

M. iddii grows up to 20 meters in height and bears white flowers.

This tree, which grows up to 20 meters (66 feet) in height and bears white flowers, is extremely rare. So far, the newly described species, belonging to the custard apple family of trees, or *Annonaceae*, is known only from the Usambara Mountains of northeastern Tanzania — a few individuals have been observed in the Amani Nature Reserve in the Eastern Usambara Mountains and one in a private reserve in western Usambara. Both reserves are 'islands' within a deforested landscape with an extensive clearance of forest in neighboring areas, the researchers write in a paper published this year. Researchers have named it *Mischogyne iddii*, after Iddi Rajabu, a resident botanist at the Amani Nature Reserve, and they estimate that fewer than 50 individuals of the tree remain in the wild.

13. A new species of venomous pit viper was described from India

The Arunachal pit viper camouflages well in leaf litter.

In May 2016, wildlife researcher Rohan Pandit and his teammate Wangchu Phiang, a member of the indigenous Bugun tribe living in the northeastern state of Arunachal Pradesh in India, were surveying Arunachal's biodiversity when they stumbled upon a snake amid the leaf litter. Pandit knew it was a species of viper, a group of venomous snakes with folding fangs, but it was unlike anything he'd seen before. So he bagged the snake and examined it in detail later, collaborating with other herpetologists to analyze the snake's morphology and DNA. The team confirmed that the viper was a new-to-science species, and they named it *Trimeresurus arunachalensis*, or Arunachal pit viper. The new species is closely related to the Tibetan pit viper (*Trimeresurus tibetanus*), a snake known only from Tibet, but physically and anatomically, the two species are quite distinct, the researchers say.

14. New species of leaf-mimicking lizard could already be victim of pet trade

Uroplatus finaritra.

The leaf-tailed gecko is a master of camouflage. These lizards, belonging to the genus *Uroplatus*, are found only in the forests of Madagascar and have body shapes and colors that allow them to merge with dried leaves seamlessly. Researchers described a new-to-science species of a leaf-tailed gecko from Marojejy National Park in northeastern Madagascar this year, and it may already be threatened with extinction because of habitat loss and the illegal pet trade. The new species, named *Uroplatus finaritra,* has a somewhat compressed body, a small triangular head, and a leaf-shaped tail, and it's a giant member of *Uroplatus*. Researchers are concerned that the species may already be a victim of the illegal pet trade since it looks similar to the satanic leaf-tailed gecko, a popular pet worldwide.

Planet Earth Is Conscious-And Life Exists in Amazing Places

15. New honeyeater species is known only from Indonesia's Alor Island

The Alor myzomela (Myzomela prawiradilagae).

This year, scientists described a new bird species that are found only on the island of Alor in eastern Indonesia. Named *Myzomela prawiradilagae* or Alor myzomela, the red-headed honeyeater is known to inhabit only eucalyptus woodland at elevations above 1,000 meters (3,300 feet) on the island, and researchers worry that its habitat on Alor is already undergoing fragmentation because of the growing human population. While locals have long known of this species, researchers hope that its description as a new species will bring in more awareness about its existence, and ensure that the species does not silently become extinct.

Planet Earth Is Conscious-And Life Exists in Amazing Places

7.0 Life Beneath and in the Antarctica Oceans

You might think that there isn't much life in Antarctica because it is mainly covered by huge Ice Sheets. But you would be wrong. Live exists in abundance under the ice in the seas and in Lakes which haven't been exposed to the rest of the Earth for many thousands of years.

7.1 Life under the Ice

A hot water drilling rig on Antarctica

Antarctica is often portrayed as a barren wasteland of ice and snow, about as inhospitable as any place on Earth. But a team of researchers just pulled up a huge amount of life from underneath the frozen continent, a testament to the tenacity of these extremophile organisms.

The life was found around 650 feet (200 meters) below the Ekström Ice Shelf, in waters that are 28 degrees Fahrenheit (minus-2 degrees Celsius) and pitch black. Seventy-seven different species of moss animals called bryozoans and worms were found, a veritable cornucopia of creatures that changes how researchers think of these extreme submarine environments. The team's research was published this week in Current Biology.

"This has massively increased the known species from this least-known habitat," David Barnes, a marine ecologist at the British Antarctic Survey, said in an email. Though some of the animals had already been found in other parts of Antarctica, the unusual habitat for this cache is a first. "This may give us clues into how life in polar seas survived glaciations," Barnes added.

The environment is difficult to access, being beneath hundreds of feet of solid ice. To actually get a look at what dwells below, the research team bored a hole through the ice using a specialized hot water drill. Then, the team dropped cameras into the borehole. They also radiocarbon dated some of the bryozoan and bivalves they found, to see how long life had been there.

"Carbon dating of dead fragments of these seafloor animals varied from current to 5,800 years," said co-author Gerhard Kuhn, an earth scientist at the Alfred Wegener Institute in Bremerhaven, Germany, in a British Antarctic Survey release. "So, despite living 3-9 kilometers [2-6 miles] from the nearest open water, an oasis of life may have existed continuously for nearly 6,000 years under the ice shelf."

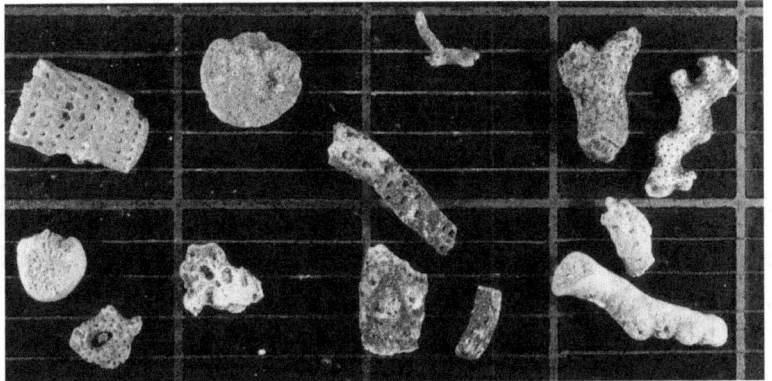
Some of the bryozoans the team found

Life persevering in such extreme conditions is impressive. Some regions under the Antarctic ice have life, despite being in complete darkness for millennia. Some microbes subsist on pulverized bedrock that settles in the sediment beneath the continent. But larger organisms also manage to get by under unthinkably challenging conditions; a different team of biologists found sponges half a mile (1 kilometer) below Antarctica's ice sheet, a discovery one of the researchers likened to "finding a bit of the rainforest in the middle of the Sahara." Though the recent team's discovery wasn't as deep, it still expands the number of environments known to sustain life.

"There are many things we can learn from this unusual (and quite large) habitat," Barnes said. "Many polar species can cope with much lower levels of food than thought, so although surface polar oceans are warming they may be able to survive in deeper (food-poor) waters."

Despite how inaccessible the habitat is, though, it is changing along with the rest of the planet. As climate change warms the planet and hastens the collapse of Antarctic ice shelves, these pitch-black inland habitats could soon be exposed to the open ocean or be altered in other ways. Even if those changes could make some locations a happier home for photosynthesizing creatures that can soak up the sun, the unique environment that currently exists under the Ekström ice shelf will be gone.

So far, only about 10 square feet (1 square meter) of the 620,000-square-mile (1.6-million-square-kilometer) habitat has actually been observed, leading to fears that some of the biodiversity under Antarctica could undergo anonymous extinction. "It is a major tragedy that one of Earth's least known, disturbed and unique habitats could be lost before we even know it," Barnes said. "There are likely to be many societally important answers to how our planet functions there."

7.2 Lake Vostok under Miles of Ice

Forget drill contaminants, anti-freeze artifacts, and human skin cells, it's finally time to bust out the most enduring quote of the nineties: Life will find a way. After much controversy and an array of scientific challenges, researchers are finally ready to confirm that life in Lake Vostok in Antarctica, which has been sealed up by four kilometers (2.5 miles) of ice for millions of years, doesn't just exist — it thrives.

Skepticism was a natural reaction when reports first started coming in about the possibility of life in the subglacial Antarctic lake. Situated under four kilometers of ice, the lake is even more inhospitable than the surface directly above it; while drilling, researchers at Vostok Station measured the air at -89 degrees Celsius, the coldest temperature ever recorded on Earth. The ice above the lake puts the water under enormous physical pressure, comparable to that found under the surface of Jupiter's moon Europa, but also cuts it off completely from the Sun. In such extreme conditions, and with essentially no outside energy input for millions of years, finding life in Lake Vostok seemed, let's just say *unlikely*.

And yet, here we are. Having found living organisms in similarly isolated caves, in the burning mouths of volcanoes, and even floating in the Earth's upper atmosphere, perhaps it shouldn't have come as such a surprise to find life in these waters. In March, researchers confirmed the existence of a genuine Vostok-dwelling bacterium. It was a species never characterized before, which was exciting, but the assumption was that this was a plucky survivor hanging on by the skin of its teeth. Now, however, it seems the species revealed a few months ago is just one of a thriving community living in this most extreme of environments.

Over 3500 different species have been identified by a form of statistical analysis known as metagenomics. Basically, they sequenced all the genetic material in their samples at once, and used sophisticated analytical techniques to make sense of the resulting jumble of information. This technique has proven extremely useful in identifying species in an agricultural soil sample, for instance, but this is the first time its use has revealed a whole group of totally novel organisms.

Planet Earth Is Conscious-And Life Exists in Amazing Places

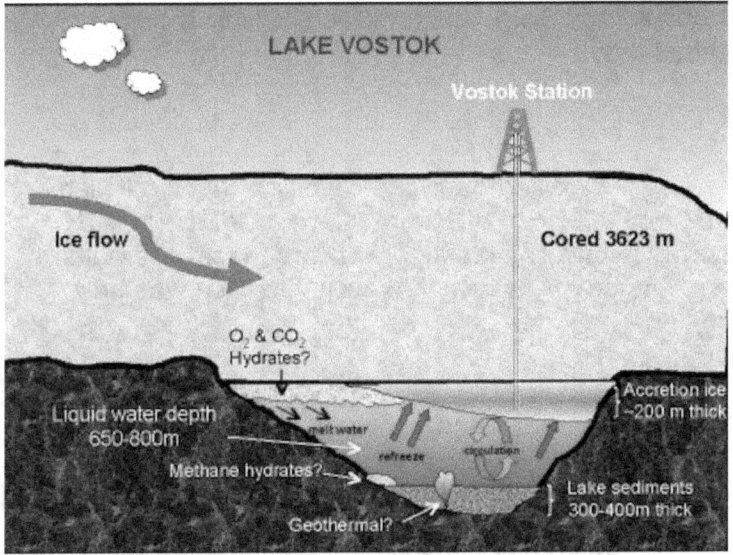

It took years to bore down to the waters they needed, but these findings were well worth the effort.

In heading off criticisms, the team addresses the idea that much of this material might be contamination or ancient life lying dead but preserved since the ice was laid down. It's impossible that the DNA could have persisted this long as a fossil, they say, but it's even *more* impossible that the RNA could have; DNA's older, more fragile sibling constitutes the best of their evidence in favor of living communities in Lake Vostok, as its half-life is far too small to be some molecular window into the past.

Most interesting is that the life they found is not entirely bacterial. Several hundred species of eukaryotic organisms also live in the water, including over 100 multicellular species. They even found species that are generally associated with mollusks and fish, leading one researcher to say that the lake "might have fish," before quickly backpedaling.

The team does note that this wide swathe of deep-lake species includes members specializing in every stage of the nitrogen and carbon cycles, implying that the isolated ecosystem of Lake Vostok may have fixed, used, and recycled its own limited supply of carbon for several million years. If true, this means that evolution has invented the ultimate sustainable green-space.

As noted, the highly pressurized conditions in of the lake are similar to those found under the ice sheet of Europa. Many have speculated that this finding could imply the existence of life on that moon, but remember that life didn't spontaneously form in Lake Vostok, but was laid there eons ago and simply managed to survive. If life exists in the underground waters of Europa, life still had to evolve on the moon at some point, in order to get down there. This finding shows, though, how long life can survive on how little energy. Could life survive a trip through space on an asteroid? Lake Vostok shows us how few environments we really can discount in our search for life in the universe.

8.0 Life in the Atmosphere

Clouds as seen from the International Space Station over Africa in 2017.

If you're feeling lonely, take solace in remembering that there are countless tiny living things floating tens of thousands of feet above your head.

And as scientists have come to learn more and more about this high-flying life and how it interacts with Earth's surface, they are beginning to question just how implausible it is to wonder whether similar life could theoretically hide out in the clouds of Venus or still more exotic worlds.

"We humans really are bottom-dwellers underneath an ocean of atmosphere above our head, and we really don't know where Earth's biosphere boundary stops at extreme altitudes," David Smith, who studies life in the atmosphere at NASA's Ames Research Center in California, said at a roundtable event held on Dec. 14 at the annual fall meeting of the American Geophysical Union, held virtually this month. "It seems just about anywhere we sample with NASA aircraft and balloons, we find signatures of microbial life."

So far, life in Earth's atmosphere seems to be strictly microbial and a temporary affair, intimately connected to life on Earth's surface rather than an independent ecosystem. Tiny, hardy organisms are swept up from the thin transition where Earth's atmosphere meets the planet and carried into the lower layers of the atmosphere on an epic detour.

"Based on what we know, the things are just moving through the atmosphere," Kevin Dillon, a Ph.D. candidate in microbiology at Rutgers University, said during the panel. "Microbes travel and use the atmosphere almost like a highway, and specifically can hitch a ride in clouds." Microbes end up in two layers of the atmosphere. In the lower troposphere, microbes mostly have to contend with the risk of drying out, Diana Gentry, a research scientist at Ames, said during the panel. Hence the appeal of clouds. "If you are picked up and suspended in the atmosphere, you're in danger of losing all of your water pretty quickly," Gentry said. "So clouds in the lower level are great — they're like these mobile water hotspots that can help keep you wet as you're picked up and transported around." In the troposphere, some microbes may survive pretty normally, even.

Meanwhile, life one level up, in the harsher stratosphere needs to contend with conditions that are still drier and even acidic. Here, microbes typically at least need to hunker down into a dormant state they can slip out of after returning to the surface. And of course, some die, and some are dead before they are swept up into the atmosphere.

Planet Earth Is Conscious-And Life Exists in Amazing Places

So far, even in the best situation, atmospheric microbes don't seem to be doing much more intriguing than simply surviving, however. "We are really just beginning to understand the dynamics of how microorganisms from Earth's surface get swept up into the atmosphere, how long they stay aloft and whether or not they're doing anything meaningful in terms of activity or growth and reproduction aloft — we still don't know," Smith said. "We have a lot more work to do in Earth's atmosphere."
But just as the discovery of hydrothermal vent communities at hot seams in the ocean floor prompted astrobiological dreams of life deep inside the hidden oceans of icy moons, so scientists now wonder whether the strange extremity of atmospheric life on Earth could be a template for determining whether anything is alive elsewhere in this chain of rocky worlds we call the solar system.

Planet Earth Is Conscious-And Life Exists in Amazing Places

9.0 Life in Isolation

Life also develops in many isolated places around and inside the Earth. Here are some examples in this chapter.

9.1 Tepui-Plateaus in South America

A tepui is a table-top mountain or mesa found in the Guiana Highlands of South America, especially in Venezuela and western Guyana. The word tepui means "house of the gods" in the native tongue of the Pemon, the indigenous people who inhabit the Gran Sabana.

Tepuis tend to be found as isolated entities rather than in connected ranges, which makes them the host of a unique array of endemic plant and animal species. Some of the most outstanding tepuis are Auyantepui, Autana, Neblina, and Mount Roraima. They are typically composed of sheer blocks of Precambrian quartz arenite sandstone that rise abruptly from the jungle, giving rise to spectacular natural scenery. Auyantepui is the source of Angel Falls, the world's tallest waterfall.

The plateaus of the tepuis are completely isolated from the ground forest, making them ecological islands. The altitude causes them to have a different climate from the ground forest. The top presents cool temperatures with frequent rainfall, while the bases of the mountains have a tropical, warm and humid climate. The isolation has led to the presence of endemic flora and fauna through evolution over millennia of a different world of animals and plants, cut off from the rest of the world by the imposing rock walls. Some tepui sinkholes contain species that have evolved in these "islands within islands" that are unique to that sinkhole. The tepuis are often referred to as the Galápagos Islands of the mainland, having a large number of unique plants and animals not found anywhere else in the world. The floors of the mesas are poor in nutrients, which has led to a rich variety of carnivorous plants, such as Drosera and most species of Heliamphora, as well as a wide variety of orchids and bromeliads. The weathered, craggy nature of the rocky ground means no layers of humus are formed.

**

Planet Earth Is Conscious-And Life Exists in Amazing Places

It has been hypothesized that endemics on tepuis represent relict fauna and flora that underwent vicariant speciation when the plateau got fragmented over geological time.

However, recent studies suggest that tepuis are not as isolated as originally believed. For example, an endemic group of treefrogs, Tepuihyla, have diverged after the tepuis were formed; that is, speciation followed colonization from the lowlands.

The tepuis, also known as 'islands above the rainforest', are a challenge for researchers, as they are home to a high number of new species that have yet to be described. A few of these mountains are cloaked by thick clouds almost the whole year round. Their surfaces could previously only be photographed by helicopter radar equipment.

Many tepuis are in the Canaima National Park in Venezuela, which has been classified as a World Heritage Site by UNESCO.

A few of the most notable of the 60 tepuis:

Auyantepui is the largest of the tepuis with a surface area of 700 km2 (270 sq mi). Angel Falls, the highest waterfall in the world, drops from a cleft in the summit.

Mount Roraima, also known as Roraima Tepui. A report by the noted South American researcher Robert Schomburgk inspired the Scottish author Arthur Conan Doyle to write his novel The Lost World about the discovery of a living prehistoric world full of dinosaurs and other primordial creatures. The borders of Venezuela, Brazil, and Guyana meet on the top.

Matawi Tepui, also known as Kukenán, because it is the source of the Kukenán River, is considered the "place of the dead" by the local Pemon Peoples. Located next to Mount Roraima in Venezuela.

Autana Tepui stands 1,300 m (4,300 ft) above the forest floor. A unique cave runs from one side of the mountain to the other.

Ptari-tepui's sheer rock walls are so isolated, it is believed a high number of endemic plant and animal species could be found there.

Sarisariñama Tepui, famous for its almost perfectly circular sinkholes that go straight down from the mountain top – the largest such sinkhole is 300 m (1,000 ft) in diameter and depth (purportedly created by groundwater erosion). They harbor an ecosystem composed of unique plant and animal species at the bottom.

Ilú-Tramen Massif is the most northerly mountain in the chain that stretches along the Venezuelan-Guyana border from Roraima in the south.

Planet Earth Is Conscious-And Life Exists in Amazing Places

Angel Falls-tallest waterfall in the world

Planet Earth Is Conscious-And Life Exists in Amazing Places

9.2 Life in Deep Caves

Here are four creepy animals that live deep down in caves:

You'll find these scary-looking creatures inside a deep, dark cave. They spend most of their lives in complete darkness. After all those years in the shadows, they've learned how to make the most of their home. Creep in and have a look!

The olm

Habitat: *the caves of Slovenia*

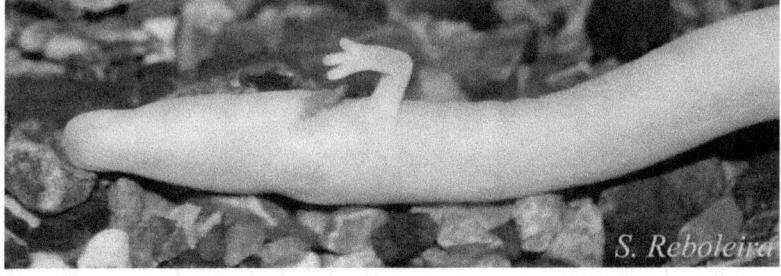

You can definitely file this cave creature under "creepy." Just looking at it is enough to make some people shiver. The *olm*, which grows to about the length of a school ruler, is a blind amphibian that looks a lot like a snake. It spends its entire life in the darkness of underwater caves. The olm relies on its powerful senses of smell and hearing to swim around its shadowy home. These super senses also help the amphibian hunt for prey, such as bugs, snails, and crabs. When it catches a meal, the olm doesn't chew it. Instead, this meat-eater swallows its prey whole. And, believe it or not, it can survive more than ten years without eating!

The cave robber spider (or *Trogloraptor*)

Habitat: *The caves of Southwestern Oregon*

A lot of people are freaked out by spiders, and when it comes to this one…you can't blame them. Its name is *Trogloraptor*, but this spider has been given the nickname the 'cave robber." This eight-legged cave dweller has sharp claws on the ends of its legs. Experts believe the creepy crawler is a fierce predator that uses those jagged claws to catch a meal. It dangles from the cave roof on a few strands of silk. When dinner gets close—snap!—the spider likely snatches up the unsuspecting prey with its claws.

The rat snake

Habitat: A cave in Yucatan, Mexico

Sssss....so spooky! You'll find most *rat snakes* in the forests of North and Central America. But one group of these snakes makes its home in a remote cave in the Mexican jungles. It turns out this cave is home to a colony of small bats. So a group of rat snakes began living in the walls and roof of the cave to feed on the bats. Locals now call it the "cave of the hanging snakes." The serpents curl up in crevices, waiting for bats to fly by. And when they do, the snakes spring into action. They dangle by their tail from the cave walls and capture the flying bats in their mouth as they zip by. Dinner is served!

the Mexican free-tailed bat

Habitat: The Bracken Cave, Texas

Planet Earth Is Conscious-And Life Exists in Amazing Places

It's the stuff of nightmares. As the sun sets, millions of bats fly out of a cave and climb up into the sky like a big, black cloud. And the bats keep coming… for three hours! That's exactly what happens every night during the summer outside one cave near San Antonio, Texas. These Mexican free-tailed bats leave their cave home each evening to feed on bugs. And the bats mean business. They gobble up tons of insects every single night! It turns out the bats live in Mexico during the winter and travel northward to roost in this one particular cave each summer. Up to 20 million bats can be found here, making it the largest colony of any bat species. With so many bats in one cave, their poop soon covers the cave floor. But it's more than just a big mess. The poop becomes a food source for tiny organisms and bugs living in the cave.

10.0 Life in Space

There is even life in Space around the Earth and probably life throughout the Cosmos. In this chapter we review bacteria which was found on the International Space Station which is new to science.

Also, there is lots of evidence of life in meteorites and I've included one meteorite story too.

10.1 Bacteria on the ISS

Previously unknown bacteria discovered on the space station could help grow plants.

After serving as a home away from home for astronauts over the last 20 years, the International Space Station has become the host for unique bacterial inhabitants. And these microbes could prove useful.

Four strains of bacteria, three of which were previously unknown to science, have been found on the space station. They may be used to help grow plants during long-term spaceflight missions in the future.

The study published Monday in the journal Frontiers in Microbiology.

The space station is a unique environment because it has been entirely isolated from Earth for years, so a multitude of experiments have been used to study what kind of bacteria is present there.

Planet Earth Is Conscious-And Life Exists in Amazing Places

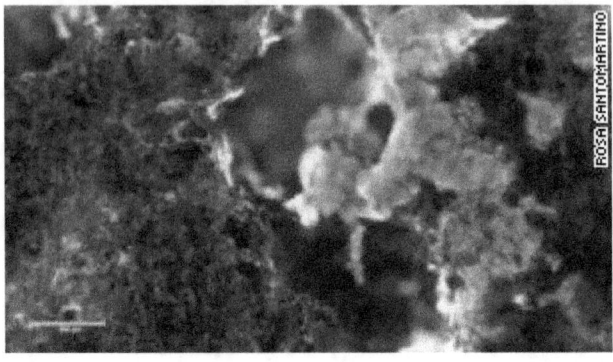

Bacteria from Earth could potentially be used to mine on the moon or Mars

Eight specific spots on the space station have been continuously checked over the last six years for the presence of microbes and bacterial growth. These areas include modules where hundreds of scientific experiments are carried out; a growth chamber where plants are cultivated; as well as places where the crew comes together for meals and other occasions.

As a result, hundreds of samples of bacteria have been collected and studied, with a thousand more waiting to return to Earth for analysis.

Planet Earth Is Conscious-And Life Exists in Amazing Places

Astronauts harvest radishes grown aboard the International Space Station

The four strains of bacteria that researchers isolated belong to the Methylobacteriaceae family. The microbes were taken from samples across the space station, during the expeditions of different crews that occurred consecutively.

Species of Methylobacterium are helpful to plants, promoting their growth and fighting pathogens that affect them, among other things.

One of the strains, Methylorubrum rhodesianum, was already known. But the other three rod-shaped bacteria were unknown -- although through genetic analysis, scientists were able to determine they were most closely related to the bacteria species Methylobacterium indicum.

The researchers want to designate the new strains of bacteria as a novel species called Methylobacterium ajmalii to honor Indian biodiversity scientist Muhammad Ajmal Khan, who died in 2019.

Senior research scientist Kasthuri Venkateswaran and planetary protection engineer Nitin Kumar Singh, both at

NASA's Jet Propulsion Laboratory in Pasadena, California, worked on this research to understand the potential applications of the bacteria.

NASA astronaut Kate Rubins collects tubes containing swab samples of microbes on the space station.
The new strains may be "biotechnologically useful genetic determinants" to assist with the growth of plants in space, the scientists said in a statement. "To grow plants in extreme places where resources are minimal, isolation of novel microbes that help to promote plant growth under stressful conditions is essential."

Leafy greens and radishes have been successfully grown on the space station, but growing crops in space is not without difficulty. Methylobacterium could be used to help plants overcome the stressors they face trying to grow outside of Earth.

However, the researchers stressed that only time, and experiments using this bacteria to test their theory, will show whether it works. The researchers also want to learn more about these newly found bacteria.

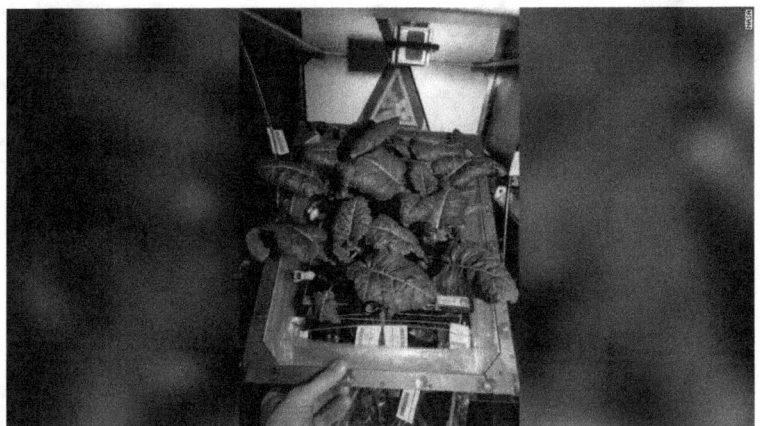

Amara mustard plants are currently being grown on the space station.

"Since these three ISS strains were isolated at different time periods and from various locations, their persistence in the ISS environment and ecological significance in the closed systems warrant further study," the authors wrote in the study.

Until humans reach Mars, the space station serves as a test bed for a multitude of technologies and resources needed for long-term missions in deep space, the researchers said. This includes the study of microorganisms and how they impact life on the space station -- and how they could be utilized.

Space-grown lettuce is safe to eat, says study. Delicious, say astronauts

"Since our group possess expertise in cultivating microorganisms from extreme niches, we have been tasked by the NASA Space Biology Program to survey the ISS for the presence and persistence of the microorganisms," Venkateswaran and Singh said.
The Methylobacterium discovered during this study isn't harmful to the astronauts, either.

"Needless to say, the ISS is a cleanly-maintained extreme environment. Crew safety is the number 1 priority and hence understanding human/plant pathogens are important, but beneficial microbes like this novel Methylobacterium ajmalii are also needed."

Given the amount of bacteria found on the space station still awaiting analysis, and the potential for discovering new strains, the researchers hope that molecular biology equipment could be developed to study the bacteria while it's on the space station.

"Instead of bringing samples back to Earth for analyses, we need an integrated microbial monitoring system that collect, process, and analyze samples in space using molecular technologies," the researchers said. "This miniaturized 'omics in space' technology -- a biosensor development -- will help NASA and other space-faring nations achieve safe and sustainable space exploration for long periods of time."

Planet Earth Is Conscious-And Life Exists in Amazing Places

10.2 Life in Meteorites

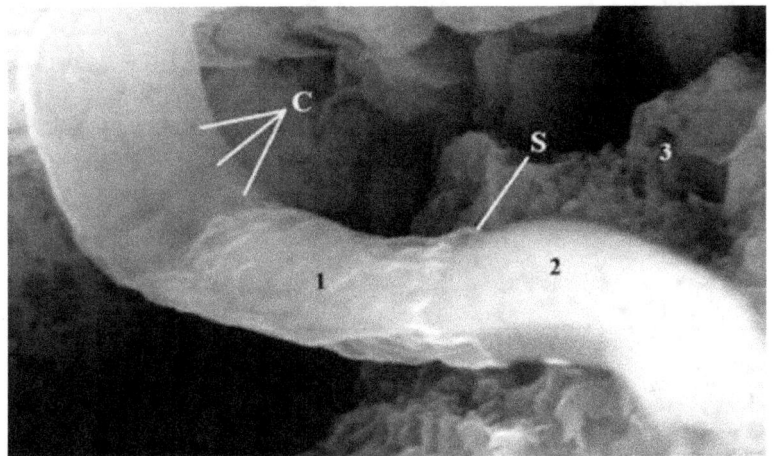

Strange life signs found on meteorites: NASA scientist

WASHINGTON (Reuters) - A NASA scientist reports detecting tiny fossilized bacteria on three meteorites, and maintains these microscopic life forms are not native to Earth.

If confirmed, this research would suggest life in the universe is widespread and life on Earth may have come from elsewhere in the solar system, riding to our planet on space rocks like comets, moons and other astral bodies.

The study, published online late Friday in The Journal of Cosmology (journalofcosmology.com), is considered so controversial it is accompanied by a statement from the journal's editor seeking other scientific comment, which is to be published starting on Monday.

The central claim of the study by astrobiologist Richard Hoover is that there is evidence of microfossils similar to cyanobacteria -- blue-green algae, also known as pond scum -- on the freshly fractured inner surfaces of three meteorites.

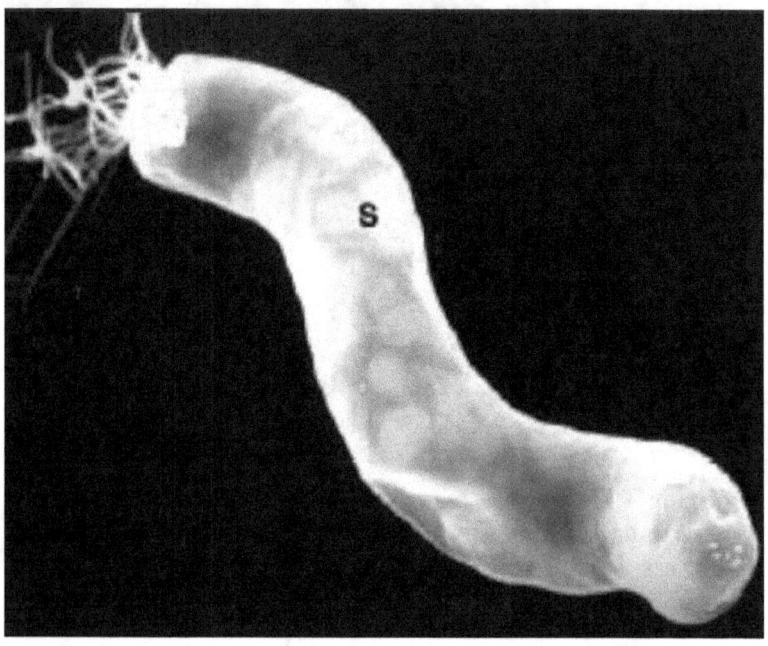

These microscopic structures had lots of carbon, a marker for Earth-type life, and almost no nitrogen, Hoover said in a telephone interview on Sunday.

Nitrogen can also be a sign of Earthly life, but the lack of it only means that whatever nitrogen was in these structures has decomposed out into a gaseous form long ago, Hoover said.

"We have known for a long time that there were very interesting biomarkers in carbonaceous meteorites and the detection of structures that are very similar ... to known terrestrial cyanobacteria is interesting in that it indicates that life is not restricted to the planet Earth," Hoover said.

Hoover, based at NASA's Marshall Space Flight Center in Alabama, has specialized in the study of microscopic lifeforms that survive extreme environments such as glaciers, permafrost, and geysers.

He is not the first to claim discovery of microscopic life from other worlds.

In 1996, NASA scientists presented research indicating a 4-billion-year-old meteorite found in Antarctica carried evidence of fossilized microbial life from Mars.

The initial discovery of the so-called Mars meteorite was greeted with acclaim and the rock unveiled at a standing room-only briefing at NASA headquarters in Washington.

Since then, however, criticism has surrounded that discovery and conclusive proof has been elusive.

Hoover's research may well meet the same fate. In a statement published with the online paper, the Journal of Cosmology's editor in chief, Rudy Schild, said in a statement:

"Dr. Richard Hoover is a highly respected scientist and astrobiologist with a prestigious record of accomplishment at NASA. Given the controversial nature of his discovery, we have invited 100 experts and have issued a general invitation to over 5,000 scientists from the scientific community to review the paper and to offer their critical analysis."

Planet Earth Is Conscious-And Life Exists in Amazing Places

11.0 Summary

Our planet Earth has much more life than scientists thought just a couple of decades ago.

The life in underground biomes going down miles underneath the surface is probably equivalent in mass to all of the life on the surface of the Earth.

Many new secret locations for life have been found in recent years, and we continue to find even more. The temperatures and pressures life can exist at now make scientists think that life can exist in other extreme environments in the Solar System.

Whether you believe that our Earth has a real consciousness or not it is clear that we live in a kaleidoscope of life of many types which surrounds us.

All the Best,

Martin K. Ettington
March 2022

Planet Earth Is Conscious-And Life Exists in Amazing Places

12.0 Bibliography

1. https://news.mongabay.com/2019/12/photos-top-15-new-species-of-2019/#:~:text=This%20year%2C%20scientists%20announced%20a,seen%20by%20scientists%20in%201914. *New life in the jungle.* [Online]

2. https://www.planetaryphilosophy.com/philosophy/philosophy-of-consciousness/the-gaia-theory/. *The Gaia Theory.* [Online]

3. https://news.fit.edu/science/the-10-weirdest-ocean-creatures-and-where-to-find-them/. *10-weirdest-ocean-creatures-and-where-to-find-them.* [Online]

4. https://gizmodo.com/scientists-found-a-cradle-of-life-under-antarctica-1848252604. *scientists-found-a-cradle-of-life-under-antarctica.* [Online]

5. https://www.space.com/earths-atmosphere-and-potential-for-life-on-venus. *Earth's Atmosphere Life.* [Online]

6. https://www.cnn.com/2021/03/16/world/international-space-station-microbes-scn-trnd/index.html#:~:text=Four%20strains%20of%20bacteria%2C%20three,the%20journal%20Frontiers%20in%20Microbiology. *Bacteria on the ISS.* [Online]

7. https://www.reuters.com/article/us-meteorites-life/strange-life-signs-found-on-meteorites-nasa-scientist-idUSTRE7252KQ20110307. *Strange Signs of Life Found on Meteorites.* [Online]

8. https://www.cbc.ca/kidscbc2/the-feed/4-creepy-animals-that-live-deep-down-in-caves. *4 Creepy Animals that live deep down in caves.* [Online]

www.ingramcontent.com/pod-product-compliance
Lightning Source LLC
Chambersburg PA
CBHW071523220526
45472CB00003B/1126